PETOSKEY STONE

Finding, Identifying, and Collecting Michigan's Most Storied Fossil

Dan R. Lynch

Adventure Publications
Cambridge, Minnesota

ACKNOWLEDGMENTS

Thank you to my wife, Julie Kirsch, for her constant support and patience throughout all of my book projects, and to my parents, Bob and Nancy Lynch, for their guidance. Additional thanks to Bob Lynch for helping find and polish specimens. I would like to also thank Bob Wright and Dean Montour for providing specimens and information.

All photos by Dan R. Lynch unless otherwise noted.

Specimens on pages 58, 59 courtesy Bob Wright

Specimens on pages 6, 47, 68, 71–72, 76–77, 81, 84 courtesy Dean Montour

Specimen on page 91 courtesy Alex Fagotti

Pictured jewelry on page 87 courtesy Bob Wright

Used under license from Shutterstock.com:
  aceshot1: 51 (top) **aga7ta:** 25 (middle) **Aprilflower7:** 25 (top) **aquapix:** 12 (inset) 14. **AuntSpray:** 23 (bottom) **buttchi 3 Sha Life:** 17 (bottom) **Cheng Wei:** 25 (bottom) **Crazy nook:** 38 (left) **Cristian Puscasu:** 30 **ehrlif:** 47 (middle), 50 **Ethan Daniels:** 29 (bottom) **Guillermo Guerao Serra:** 22 (right) **Kenneth Sponsler:** 46, 47 (top) **Marius Meyer:** 16 **Michael Deemer:** 55 (top) **mushika:** 34 **Narongsak Nagadhana:** 40 **Nicks Wanderings:** 24 **Northern Way of Life:** 53 (bottom) **Nuno Vasco Rodrigues:** 15 (bottom left) **Richard Whitcombe:** 29 (top) **scubaluna:** 15 (bottom right) **SeraphP:** 13 (bottom) **SL-Photography:** 31 **smspsy:** 26 **unterwegs:** 17 (top) **UWPhotog:** 12 (main), 13 (top)

All illustrations and maps by Dan R. Lynch

Cover and book design by Jonathan Norberg

PETOSKEY STONE

Finding, Identifying, and Collecting
Michigan's Most Storied Fossil

TABLE OF CONTENTS

THE MICHIGAN STATE STONE

Since June 28, 1965, a peculiar brown rock has represented Michigan as its state stone. Called Petoskey stone, this fossil-bearing limestone represents the rich history of ancient Earth as much as it does the identity of modern-day Michigan. Each pebble, found in the waves and sand on the shores of Lake Michigan and Lake Huron, is a relic from an Earth that we wouldn't recognize: an Earth with seas that teemed with life, while little more than fungi and primitive plants ruled the land. But the coral preserved in today's Petoskey stone didn't merely just exist during this period of time—it helped define it, contributing to the enormous coral reefs that formed the backbone of the oceans. These early reefs were a haven for life, places where evolution could occur relatively swiftly, and the coral in Petoskey stone not only helped to support much of the burgeoning marine ecosystem during this critical point in history but also the development of life as we know it. While Petoskey stone is rightly famous as a collectible, many who seek it today may not realize how it contributed to making the Earth a more hospitable place for billions of species that followed it, including us.

PETOSKEY STONES, AND HOW THEY FORMED

Petoskey stones have long been easy to find in Michigan. The Little Traverse Bay, where the highest concentration of Petoskey stones are found, is a beautiful natural area that has drawn in collectors for many years and offered them beautiful water-worn specimens of their own.

But while finding them may be a fairly simple matter, answering the question of "what is a Petoskey stone?" is a little more complex. With over 400 million years of history, there's a lot to know about the little hexagons trapped in the rock, and it all starts in the ancient seas of Earth. In this section, we'll explore the deep history of this enigmatic stone and how it came to be.

Where Petoskey stones are found in Michigan

Specimen courtesy of Dean Montour

WHAT IS PETOSKEY STONE?

Petoskey stone is undoubtedly Michigan's best known fossil, famous for its enigmatic mosaic-like pattern caused by the remains of an ancient coral species preserved in limestone. It is found most often in the vicinity of the Little Traverse Bay Area, particularly around Charlevoix and the town of Petoskey, from which the stone gets its name, but the popular collectible can be found all across the northern portion of Michigan's Lower Peninsula.

Until recently, the precise identity of the coral species preserved in Petoskey stone was unclear. For some time, scientists have known that Petoskey stone is a fossil remnant of corals belonging to the genus *Hexagonaria*, a long-extinct group of corals that developed hexagonal, or six-sided, patterns. These corals thrived in Earth's shallow seas during the Devonian Period, which started approximately 419 million years ago. But it wasn't until 1969 that the primary species preserved in the rock was identified as *Hexagonaria percarinata*, an ancient coral found in rocks in several places in the world, but nowhere as plentiful as in Michigan. Appearing tightly packed together, the hexagonal shapes each represent an individual organism within the once-living coral colony. Each hexagon is segmented, has a gauze-like texture, and often has a central darker spot.

Hexagonaria percarinata, typically shortened to *H. percarinata*, is a scientific name that will be repeated throughout this book; a coral fossil is only an "official" Petoskey stone if it contains that specific species. But that distinction isn't important to every collector, as there are many corals in the *Hexagonaria* group, and most *Hexagonaria* fossils from the region share similar patterns and features and formed in the same rock bodies. Instead, to many people, any fossil find is exciting, as they all represent the rich and deep history of life in Michigan and make for beautiful keepsakes.

Most Petoskey stones are found loose in surface gravel or sand; they're found as weathered, rounded stones. Less weathered specimens appear more ragged and rough, and these better represent the coral as it originally grew.

PETOSKEY WITHIN SOUTHERN MICHIGAN

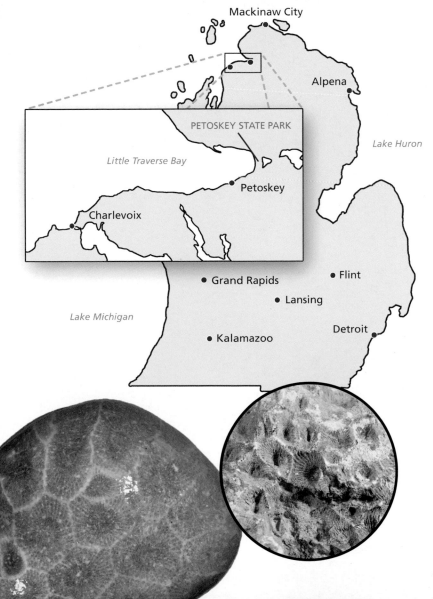

Mackinaw City

Alpena

Lake Huron

PETOSKEY STATE PARK

Little Traverse Bay

Petoskey

Charlevoix

Grand Rapids

Flint

Lansing

Lake Michigan

Detroit

Kalamazoo

Petoskey stones are known for their tightly packed, repeating patterns of hexagons. Like most other examples of limestone, they are usually brown, tan, or yellowish, and fairly soft, yet will take a good polish, as shown by this exemplary specimen.

Most Petoskey stones are found on beaches where they have been worn down and smoothed by the wind and waves. These specimens are classic examples of the kind you'll easily find on the shore. Note how subtle the pattern can be in very worn specimens.

When cut, polished, and viewed under magnification, the complexity of the coral's shape and structure can be fully appreciated.

Not all Petoskey stones are found on the shore. Inland finds tend to be less weathered and can exhibit more detail, as well as reveal some of the original cup-like shapes of the coral surface.

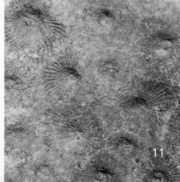

PETOSKEY STONE IS A FOSSIL CORAL

To understand Petoskey stone, you need to know about the animals that created it: corals. Both ancient and modern corals share a lot in common. What's more, corals alive today can shed light on how ancient corals grew and lived, and studies of fossil corals have revealed evidence of past environmental collapses that may be relevant today.

Corals are ancient invertebrate animals, first appearing in the fossil record during the Cambrian Period, which began 542 million years ago. They are still alive today and are among the longest-living groups of animals on record. Both ancient and modern corals are small organisms that live in marine environments, usually in large, tightly compact colonies. Corals are best known for their ability to secrete external skeletons composed of **calcium carbonate**, a compound consisting of calcium, carbon, and oxygen. This means that the branching, bulbous, or fan-like shapes we know as "coral" actually consist of thousands to millions of individual organisms. These are coral colonies, which are housed in a collective coral skeleton.

Modern corals in a reef

Corals are a vital part of ocean ecosystems around the world because many coral species help build **reefs**, large stony masses that form after successive generations of coral growth.

Reefs are particularly important because corals thrive in waters that are low in nutrients. And while corals rely on ocean currents to carry prey to them, the reefs they form provide shelter and food sources for countless species of fish, mollusks, cephalopods, crustaceans, sponges, and many other kinds of animals. The resources that corals and reefs provide other ocean organisms are critical for seas to thrive. The species of coral in Petoskey stone, *H. percarinata*, played a similar role in the ancient oceans.

A healthy reef

CORAL ANATOMY

The individuals within a coral colony are called **polyps**, and each polyp lives in a small circular depression, called a **corallite**, in the surface of a hard coral structure. Polyps are typically very small, usually more-or-less cylindrical in shape, and fleshy. They are topped with a circular array of tentacles that catch their prey—usually plankton, but certain species feed on small fish as well. The tentacles direct prey toward the coral's centralized mouth. At the base of each polyp are specialized structures that secrete calcium carbonate (in the form of aragonite or calcite, depending on species). The calcium carbonate hardens to build the trademark coral structures of coral reefs.

Between each polyp is a living tissue called the **coenosarc** (pronounced seen-oh-sark); it coats the outer surfaces of the stony coral colonies and connects each individual polyp. It also secretes its own calcium carbonate, contributing to the hard reef structure. This elaborate network of living tissue allows the entire colony to share nutrients and ensures the survival of the colony as a whole. It also provides a means of chemical communication, aiding in reproduction. Most corals reproduce sexually, with colonies releasing both eggs and sperm into the water en masse during simultaneous spawning events, which are usually triggered around full moons.

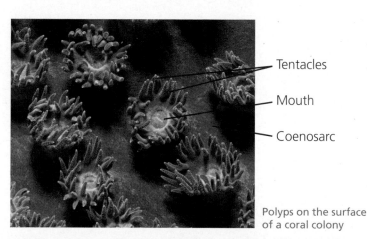

Tentacles

Mouth

Coenosarc

Polyps on the surface of a coral colony

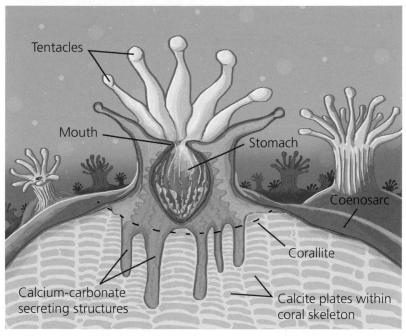

Coral polyp in corallite (cross section)

Modern coral polyps

OTHER LIFEFORMS IN CORAL

Coral polyps aren't the only organisms living within a coral skeleton. Like many other ocean species, coral polyps have evolved a special relationship with tiny single-celled organisms called zooxanthellae. **Zooxanthellae** (pronounced zo-ah-zan-thel-ay) live within the polyps themselves, something like an infection, but they do not harm the polyps. Instead, both organisms help each other thrive; in biology, this is known as symbiosis.

Zooxanthellae primarily produce energy via photosynthesis, and they obtain nutrients from the waste products that coral polyps produce. Similarly, the zooxanthellae produce oxygen and glucose as waste, and the coral polyps use these nutrients for energy. This unique relationship keeps both organisms healthy, and in fact even gives corals their bright colors—corals are naturally white, and those with healthy zooxanthellae populations have more vibrant colors. Conversely, corals in polluted or overly nutrient-rich (eutrophic) water eject the zooxanthellae as an emergency measure in order to conserve their own energy. This results in coral bleaching because the coral loses its color in the process. If the water quality returns to an ideal state, they can reacquire zooxanthellae, but in many cases, the bleached coral starves and dies.

Corals get their coloration from the zooxanthellae living within them, as vividly illustrated by this modern Dragon Eye Coral.

Tightly packed living corallites like these are a good example of how the Petoskey stone's patterns came to be.

Healthy coral with vibrantly colored zooxanthellae and lively polyps

Coral that has lost its zooxanthellae "bleaches" to a sickly white coloration. This coral will soon starve and die as a result of its water becoming polluted and too warm.

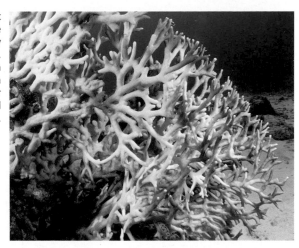

Zooxanthellae did not yet exist in the Devonian Period, when *Hexagonaria* lived, but it is thought that most corals throughout history have shared a similar symbiotic relationship with other single-celled organisms and algae. As a result, *Hexagonaria* corals may have been vibrantly colored as well.

HEXAGONARIA AND OTHER RUGOSE CORALS

Hexagonaria percarinata (or *H. percarinata*, for short) is the coral present in Petoskey stone. It was a **rugose coral**, a variety of coral with a rugose, or "wrinkled," exterior wall texture. While *Hexagonaria percarinata* was colonial, forming tightly packed groups of individuals, many rugose corals were solitary, with each polyp growing as an individual tube-like structure. These often resembled a curved, pointed horn, hence their nickname, "horn corals." All rugose corals are now extinct, but their fossils are among the most common invertebrates found today, especially in Michigan's limestone formations.

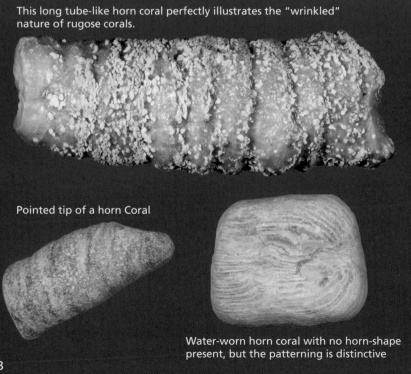

This long tube-like horn coral perfectly illustrates the "wrinkled" nature of rugose corals.

Pointed tip of a horn Coral

Water-worn horn coral with no horn-shape present, but the patterning is distinctive

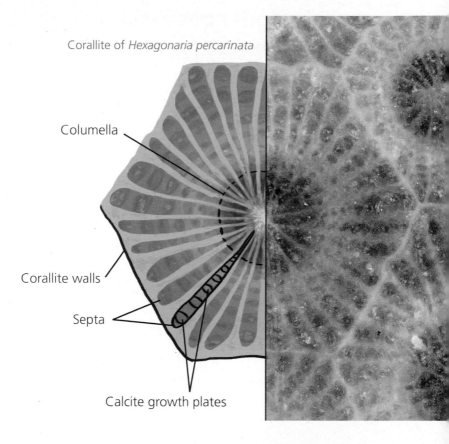

Corallite of *Hexagonaria percarinata*

Columella

Corallite walls

Septa

Calcite growth plates

H. percarinata was also a **stony coral**, producing large, hard dome-like structures composed of hard calcite. These domes often began small, sometimes with just dozens of corallites, but could grow to enormous sizes and contain tens of thousands or even millions of individuals, producing immense coral reefs on a scale unlike anything that exists today. The corallites themselves are taller than they are wide and contain vertical segments, called **septa**, that are supported by a central rod-like structure called the **columella**.

THE LIFE STORY OF THE PETOSKEY STONE

As beautiful and interesting as a polished Petoskey stone specimen may be, the story of how its coral came to be preserved in Michigan's limestone is even more compelling. It begins during the Devonian, a time period that began around 419 million years ago and lasted the next 60 million years. Known as "The Age of Fish," the Devonian world was very different from ours, with the continents closer together and inundated by warm, shallow seas that covered much of the planet, including modern-day Michigan. And, true to its nickname, the Devonian era saw the evolution of an incredible diversity of fish and other sea life—including coral—for the first time in Earth's history.

Earth in the Early Devonian Period
(and the approximate locations of modern-day continents)

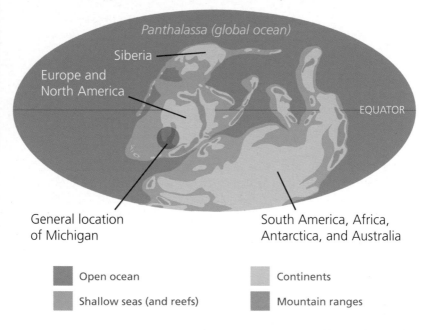

Panthalassa (global ocean)

Siberia

Europe and North America

EQUATOR

General location of Michigan

South America, Africa, Antarctica, and Australia

Open ocean		Continents	
Shallow seas (and reefs)		Mountain ranges	

Earth's Geological Periods (and when they started)

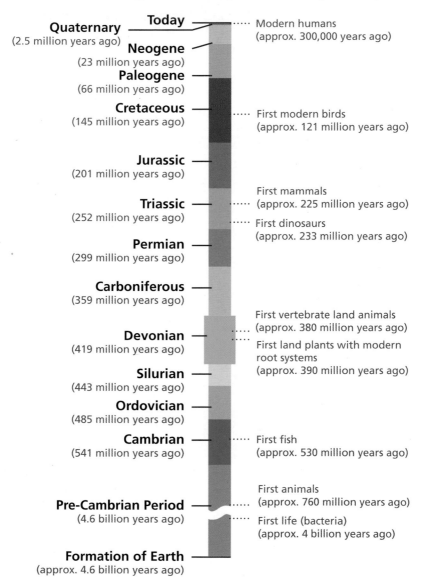

Today

Quaternary
(2.5 million years ago)

Modern humans
(approx. 300,000 years ago)

Neogene
(23 million years ago)

Paleogene
(66 million years ago)

Cretaceous
(145 million years ago)

First modern birds
(approx. 121 million years ago)

Jurassic
(201 million years ago)

Triassic
(252 million years ago)

First mammals
(approx. 225 million years ago)

First dinosaurs
(approx. 233 million years ago)

Permian
(299 million years ago)

Carboniferous
(359 million years ago)

First vertebrate land animals
(approx. 380 million years ago)

Devonian
(419 million years ago)

First land plants with modern
root systems
(approx. 390 million years ago)

Silurian
(443 million years ago)

Ordovician
(485 million years ago)

Cambrian
(541 million years ago)

First fish
(approx. 530 million years ago)

First animals
(approx. 760 million years ago)

Pre-Cambrian Period
(4.6 billion years ago)

First life (bacteria)
(approx. 4 billion years ago)

Formation of Earth
(approx. 4.6 billion years ago)

LIFE IN THE DEVONIAN

The Devonian was also a relatively warm time, and glaciers likely didn't exist for most of the period. As a result, sea levels were much higher, with less dry land around the globe than we are familiar with today. In fact, much of the Earth was primarily covered by one enormous ocean, named Panthalassa. This gave sea life the opportunity it needed to flourish and explode in diversity and size. The ocean lifeforms that were common in previous geological eras, such as trilobites and crinoids, still existed in the Devonian, but vertebrates, especially jawed fish, rose to become a dominant force.

Bony fishes and sharks were prevalent in Devonian seas, as were placoderms, which were essentially armored fish. Placoderms had bony plates covering their heads and parts of their bodies, and several species reached many feet in size, though all are extinct today. Countless other fish species, many quite unlike any alive today, saw population booms, and in general, fish diversity and complexity exploded. Many of these species were able to thrive in part due to the enormous coral reefs that formed off the coasts, to which *H. percarinata* was an important contributor.

Crinoids, nicknamed "sea lilies," may look plant-like, but they are remains of anchored fan-like animals. Their "stems" were segmented and often fell apart when they died. As a result, small disc-like fragments are very common fossils.

Trilobites ruled the sea floor during the Devonian and were highly successful animals; today, they are extinct.

Though this particular specimen is from the Silurian period, eurypterids like this were a dominant force in Devonian seas.

An artist's rendition of what Earth looked like during the Middle Devonian. The fish at the center is an armored placoderm feeding upon a eurypterid, also known as a "sea scorpion." Trilobites can be seen along the sea floor while ferns and other early plants dominate the land.

DEVONIAN REEFS AND FORESTS

Devonian coral reefs were extensive, fringing islands and even continents. They were constructed jointly by rugose corals, such as *Hexagonaria*, calcerous algae (algae that have calcium carbonate anchors), and especially by stromatoporoids, sea sponges that build coral-like structures. It was in these reefs that *Hexagonaria* grew its trademark six-sided corallites, tightly packed in dome-like coral skeletons.

On land, life was just getting its start at the beginning of the Devonian. The world would be largely unrecognizable to us today, as only spongy mats of bacteria, algae, and mosses dominated the land. Since these "plants" lacked roots and any real way of supporting themselves, the earliest "forests" were only inches high. But that was more than enough for the first animals that roamed them: tiny creatures, particularly arthropods such as mites and early scorpions. It wasn't until the Middle Devonian, several million years later, that the first insects began to appear, alongside plants with roots—most notably ferns and horsetails. Rooted plants began to transform the landscape and were able to spread much more widely than their rootless predecessors. It didn't take long for vast forests of shrub-height to cover the Earth, followed soon after by the first trees. This period of rapid plant growth is known today as the "greening of the Earth" because the dry surfaces of the planet went from being largely barren rock to plant-covered in a relatively short amount of time. By the Late Devonian, the first seed-bearing plants appeared—before then, plants reproduced only by releasing spores, like ferns, or by vegetative growth.

Mats of algae, like this on some waterside rocks, would have been extremely common in the Devonian.

Ferns and horsetails were two of the primary land plant groups in the Middle Devonian.

Algal mats were a regular part of the landscape during the Devonian, both on land and in the water.

FATAL ALGAL BLOOMS

In the Devonian, the oceans and landmasses were both thriving for the first time in Earth's history. In the Early Devonian, the land was bare, and what little life existed on land had little effect on the oceans. But as plants began to take over the land, their roots broke up rocks and created incredible amounts of soil for the first time. The presence of soil further erodes rock, which in turn releases minerals and ions that algae and plants use as nutrients. Since this had not occurred on Earth before the Devonian, sea life had evolved in the relative absence of such nutrients. But as soil developed quickly, the explosion of plant life across the globe resulted in huge amounts of nutrient-rich runoff entering waterways and oceans for the first time.

The rapid introduction of so many nutrients likely caused great blooms of algae in rivers, lakes, and the oceans. Algae use carbon dioxide to grow, producing oxygen as a waste product. But when algae die, they also consume oxygen; as algae break down, their carbon combines with oxygen to create carbon dioxide. This is key because, as the ancient algae died and sank, much of the oxygen in the water was used up in the process, and the resulting carbon dioxide then contributed to further algae growth. Due to the massive amounts of algae, the rate of decomposition likely exceeded the rate of oxygen production, thereby creating anoxic (oxygen-depleted) conditions that would have suffocated fish and other oxygen-dependent life, including *H. percarinata*. This process, called eutrophication, was a catastrophe for ocean life in the Devonian, and it was all caused by the rapid expansion of plant life on land.

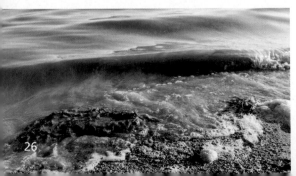

Algal blooms can be as devastating to aquatic environments today as they were in the Devonian. The enormous amount of algae simply chokes out other sea life.

Nutrient-rich Runoff Causing Catastrophic Algal Blooms

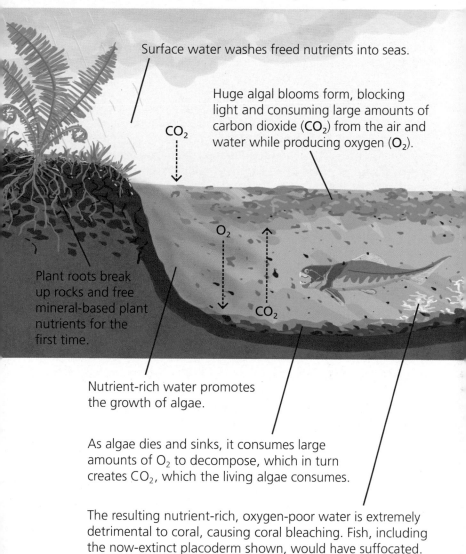

Surface water washes freed nutrients into seas.

Huge algal blooms form, blocking light and consuming large amounts of carbon dioxide (CO_2) from the air and water while producing oxygen (O_2).

CO_2

O_2

CO_2

Plant roots break up rocks and free mineral-based plant nutrients for the first time.

Nutrient-rich water promotes the growth of algae.

As algae dies and sinks, it consumes large amounts of O_2 to decompose, which in turn creates CO_2, which the living algae consumes.

The resulting nutrient-rich, oxygen-poor water is extremely detrimental to coral, causing coral bleaching. Fish, including the now-extinct placoderm shown, would have suffocated.

THE LATE DEVONIAN EXTINCTION

Fish and other vertebrates weren't the only sea life negatively affected by the greening of the Earth. Corals, which best thrive in nutrient-poor water, were suddenly competing with growths of algae and other life that made reef production difficult.

Nutrient-rich runoff has severe detrimental effects on coral reefs, as shown today by what is occurring in Australia's Great Barrier Reef. Farm field runoff rich in phosphate fertilizers is causing the corals of the Great Barrier Reef to eject their zooxanthellae, resulting in mass coral bleaching. As the coral starves and dies, algae take over and reef-dwelling fish leave the area, thereby eliminating food sources for other animals. This disturbance travels up the food chain, affecting sea turtles, birds, and humans. A similar process very likely took place during the Late Devonian Extinction and would have devastated reef habitats, where huge numbers of species depended on the reefs even more than they do today.

What's more, the exploding numbers of plants and algae would have absorbed more carbon dioxide than ever before in Earth's history. Since carbon dioxide is a greenhouse gas that helps the Earth's atmosphere retain heat, the sudden reduction of carbon dioxide likely led Earth's global temperature to decline. This would have been a dramatic problem for many animals and plants, which for millions of years had only known a warmer Earth. The cooling of the oceans and the "over-fertilization" of the water is thought to be the primary cause of a massive extinction event, called the **Late Devonian Extinction**, in which a vast number of animal groups perished. The fossil record shows that this extinction event affected warmwater organisms, such as corals, more than any other group of animals. This included all species of *Hexagonaria*, which began to disappear from the fossil record during the Late Devonian Extinction.

Masses of dead, bleached coral are becoming all too common around the world today, especially at the Great Barrier Reef. As the reefs die, algae take over, and fish and other animals leave the area.

OTHER FACTORS IN THE EXTINCTION

Dramatic algal blooms were likely not the only factor in the Late Devonian extinction. Massive volcanic eruptions in what is now Siberia may have played a role as well. Such eruptions would have released enormous amounts of gases that would have blocked sunlight, further lowering temperature. No matter the specific cause of the extinction, we know that the Earth cooled significantly during the Devonian as we have geological evidence showing that glaciers first appeared late in the period. This caused sea levels to drop, further exacerbating the crisis in the oceans.

All of these events—choking algal blooms, cooler global temperatures, dropping sea levels, possible toxic gases in the atmosphere—spelled doom for the oceans. Countless marine species were unable to adapt and began to die off in the Late Devonian Extinction event. *Hexagonaria* species, including the one that would produce the Petoskey stone fossil, disappeared during or shortly after the extinction. This massive die-off is part of the reason why so many *Hexagonaria* and other fossils are present in Devonian rocks around the world today. And while many rugose corals survived the extinction, the final remnants of the coral group eventually went extinct 120 million years later during the greatest extinction event of all: the End-Permian Extinction, which wiped out 95 percent of all ocean species.

Some of the greatest environmental calamities in history were caused by large emissions of volcanic gases, and they may have contributed to the Late Devonian extinction event. In this photo, volcanic gases billow from the Earth.

HOW OLD ARE MICHIGAN'S PETOSKEY STONES?

Michigan's Petoskey stone was deposited in the Middle Devonian (approximately 393 million years ago), well prior to the Late Devonian extinction event. This makes Petoskey stones approximately 393 to 382 million years old. We know from other rock deposits around the world that *H. percarinata* and other *Hexagonaria* species survived until (and perhaps slightly beyond) the Late Devonian extinction, but Michigan's Petoskey stone is not from that turbulent period.

The fossils that produced Petoskey stones were deposited in the Middle Devonian, as younger coral populations boomed, burying the older remains. As sediments filled in around the *H. percarinata* skeletons, they were preserved within freshly formed limestone. As the rock solidified, the limestone remained deeply buried and protected from weathering, allowing fine details to remain preserved.

Huge vents like this pour enormous amounts of toxic gasses into the atmosphere and can contribute to global cooling.

FOSSILIZATION

Most of what we know about ancient life was learned from studying fossils. A fossil forms when the tissues of a once-living organism, whether a plant or an animal, are replaced with minerals. Usually, this occurs when the organism is buried or sinks in an environment with little oxygen, which slows down decomposition. If buried completely, the tissue may never fully decompose, leaving it trapped in sediment. As the sediment turns to rock, dissolved minerals can penetrate and replace the organic material with calcite, quartz, or other minerals. In other cases, carbon released from the organic matter, in the form of carbon dioxide, reacts with minerals in the surrounding rock, such as iron, and replaces the organic matter with minerals that the decomposing body helped to create. In either case, the mineralized material perfectly preserves the animals and plants in which it formed, and the fossil's "new" material is often quite sturdy, enabling it to endure eons within rock.

In the case of corals, however, much of the "work" of fossilization is already completed while the colony is still alive. Because corals already create hard, mineralized skeleton structures themselves, corals don't need mineral replacement to be preserved. Instead, since coral reefs naturally grow upward upon older reef material below, coral can become a fossil simply by remaining intact long enough for other sediment to fill in around it. Eventually, the buried parts of the reef turn into limestone, a type of sedimentary rock that consists entirely of fossilized marine organisms.

This fragmented mass of limestone, found in Charlevoix alongside Petoskey stones, contains traces of fossil matter but no well-developed coral.

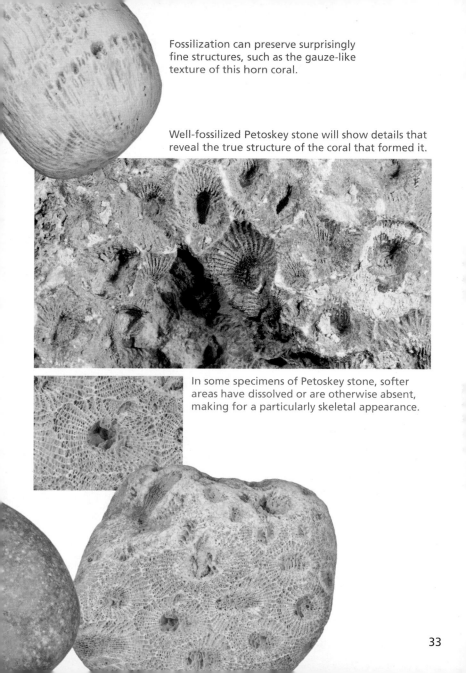

Fossilization can preserve surprisingly fine structures, such as the gauze-like texture of this horn coral.

Well-fossilized Petoskey stone will show details that reveal the true structure of the coral that formed it.

In some specimens of Petoskey stone, softer areas have dissolved or are otherwise absent, making for a particularly skeletal appearance.

33

LIMESTONE

Petoskey stone is a variety of limestone. While there are lots of fossil-bearing rocks in southern Michigan, only limestone will hold *H. percarinata*, so learning to identify it and tell it apart from other rocks will be crucial to finding Petoskey stones. Most samples of limestone will look very similar, but not all of them will contain the telltale patterns of fossil corals.

Limestone is a common sedimentary rock; as the name indicates, sedimentary rock consists of sediments, or particles, of other material. The sediments found in most limestone consist almost entirely of fossil matter, mainly the hard skeletons of marine invertebrates. Corals and reef materials are a primary component of limestone, as are the hard calcium carbonate shells of mollusks (clams and snails) and foraminifera (tiny organisms that live in sea-floor sediments). As these skeletons and shells continued to build up into thick beds on the sea floor, they began to compact and, as time went on, the calcium carbonate sediments began to harden together, solidifying the material into a layer of rock. As such, limestone itself can be thought of as a fossilized reef, whether any discernible fossils are visible or not.

The tiny shells of foraminifera are a major component of limestone and an abundant form of sea life still present today. The charming "star sand" of Okinawa, Japan, is comprised entirely of calcium carbonate foraminifera shells.

Fossil-bearing limestone from the Little Traverse Bay Area

Other Michigan limestones

Process of Reef Build-up and Limestone Production

Coral grows and builds upon itself, thickest just offshore

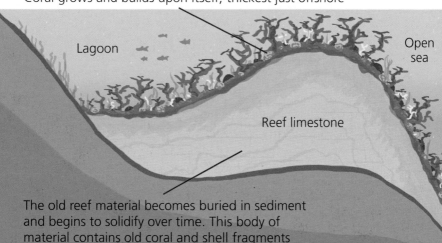

Lagoon

Open sea

Reef limestone

The old reef material becomes buried in sediment and begins to solidify over time. This body of material contains old coral and shell fragments from countless different animals.

Bedrock

THE MINERALS THAT MAKE UP LIMESTONE

The fossils that make up limestone—marine shells and skeletons—consist of calcium carbonate, a chemical compound that occurs as one of two minerals: calcite or aragonite, depending on the animal that produced it. Most mollusks, such as clams, build their shells of aragonite, but some corals, including *H. percarinata,* produced their structures from calcite. Both are common minerals made up of the elements calcium, carbon, and oxygen; the key difference between them lies in their crystal structure. The molecular structure of aragonite is less stable than that of calcite, and as a result, aragonite slowly alters, or transforms, into calcite (but usually with little to no visual change). Limestone can also contain minor amounts of other minerals, namely dolomite, quartz, clay minerals, and even other rock fragments, such as sand.

HOW TO IDENTIFY LIMESTONE

Limestone tends to be light colored, usually white or gray to yellow, tan, or brown, but other minerals present within it can tint it other colors. Sometimes it can also be colored very darkly, even black, which is often caused by higher concentrations of organic matter.

When it comes to texture, limestone is usually very fine-grained, with individual calcite grains visible under magnification and often giving a freshly broken specimen a slightly "glittery" appearance. When freshly exposed or broken, most limestones exhibit a rough surface with fairly sharp edges, sometimes with indications of flakiness. As limestone weathers, however, it becomes smoother and duller, and its grains (and any colored layers or patterns it may have) become less defined. When limestone is weathered, it can look and feel particularly chalky or dusty in your hands.

Many limestone formations are also riddled with small cavities or holes, called vugs, which may by lined with crystals, such as those of calcite, aragonite, dolomite, or a number of other minerals. Vugs often formed when larger fossils trapped in the rock dissolved, leaving behind a cavity.

These close-ups show the texture of limestone.

Calcite crystals: Calcite is the primary constituent of limestone.

Aragonite crystals: Despite their identical composition, it's easy to see how aragonite's crystal structure differs from calcite's.

Many limestones are evenly colored without much visual interest, but finds bearing larger, more visible fossils are a different story. Dramatic fossils, like dinosaur bones and ancient fish skeletons, are rare, but small curving shapes, circular disks, gauze-like textures, and ringed patterns of dimples are all subtle but important and common signs of ancient life. Learning to interpret some of these shapes can take some in-depth research, though. *Hexagonaria percarinata* is a prime example; the often-subtle honeycomb-like pattern of a rough specimen would likely go unnoticed or be misinterpreted if Petoskey stone were not already so well known.

When weathered, indistinctly colored, or lacking any visible fossils, limestone can be tricky to identify, despite its abundance. Thankfully there are a few easy tests you can perform to identify it. Limestone isn't a very hard rock; a pocket knife or a piece of glass will easily scratch it. In addition, any acid, including weak household acids such as vinegar, will make limestone effervesce, or fizz, and produce small bubbles as it dissolves. Stronger acids, such as undiluted vinegar, will elicit a stronger reaction but aren't necessary to see a result. These tests, along with color and abundance, are usually all that's needed to identify it. These traits will all help you distinguish the Little Traverse Bay area's limestone from shale and sandstone, the other prominent rocks in the area, which do not contain *H. percarinata*. Spotting limestone will help you stay on track to finding Petoskey stones of your own.

 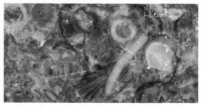

The left photo shows modern day reef sediments consisting almost entirely of various shell and coral fragments, and the right photo shows a fossil limestone containing large amounts of similar ancient sea-life fragments. Though not from the Devonian Period, this fossil-rich rock perfectly illustrates how limestone can be thought of as a fossilized reef.

Vug in limestone

Many of the most abundant limestone fossils are somewhat faint, such as this large horn coral "ring."

Travertine, a form of chemically deposited limestone that contains no fossils

As a final note, while most limestones are formed of fossil matter from marine environments, it should be noted that some less common varieties of limestone were formed chemically from the precipitation of calcium carbonate dissolved in groundwater. The most notable varieties formed this way are travertine and onyx, typically deposited around hot springs. These contain no fossils nor even the tiny grains characteristic of most limestone; instead they are made up of layers of crystallized calcite or aragonite.

ICE AGES AND GLACIERS

The body of rock containing Petoskey stone remained deeply buried for millions of years beyond the Devonian, and it wasn't until Earth's last glacial period, beginning approximately 115,000 years ago and lasting until just around 11,000 years ago, that the Petoskey stone-bearing rock layers were uncovered. Glaciers are immense sheets of slow-moving ice; during the last ice age, they descended from Canada, and their incredible size and weight scoured and bulldozed the landscape. Harder rocks were worn down, while soft rocks, especially limestone, were pulverized. But as deeper rock layers were revealed and subsequently broken up, fragments ranging in size from boulders to pebbles were pushed and scattered all over. This included the layers containing Petoskey stones.

Further weathering in wind, waves, and ice continued to round and smooth many specimens. Others were freed from their host rock only to be again buried in gravel created by the glaciers, which presents us with specimens that are still fairly easy to collect but retain more better-preserved coral features. And, due to the sheer abundance of fossil corals in the region's Devonian rocks, countless Petoskey stones have already been collected, but countless more are still waiting to be discovered.

As this glacier recedes, it leaves behind large amounts of gravel and boulders originally plucked and scraped from the underlying bedrock.

Glacial Scouring, Plucking, and Dumping

Movement of glacier

Rock fragments are plucked, or pulled, off the surface of bedrock as the glacier passes over.

The weight of the ice and the scouring action of the rock fragments trapped in the ice pulverize soft rocks and create gravel and sand. Basins and valleys, like those that hold today's lakes and rivers, are also created.

As the glacier terminates, or ends, and eventually recedes, the sand and gravel contained within it is dumped in large piles, called glacial till. Collectibles like Petoskey stone often end up here.

Rounded, smoothed Petoskey stones like these were largely shaped by the destructive action of the last ice age's glaciers.

PETOSKEY STONES TODAY

While *Hexagonaria percarinata* fossils can be found in different bodies of rock throughout the world, Michigan's Petoskey stone only originates from a band of limestone deposited in the middle-Devonian. As sea-floor silt buried the lower portions of a reef, the corals were preserved in the resultant limestone and buried beneath layer upon layer of later rocks, where they remained for millennia until the most recent glaciers revealed them.

Petoskey stone largely formed within a particular layer of sedimentary rocks called the Gravel Point Formation, itself a part of a larger group of rock layers known as the **Traverse Group**, all deposited as a result of settling ocean sediments during the Middle Devonian. The Traverse Group is composed largely of limestone, but also of layers of shale; fossils of many different kinds are prevalent within all of the group's rocks.

Common beach-worn limestone pebbles

Rough, layered chunks of shale

Southern Michigan's Devonian Rocks

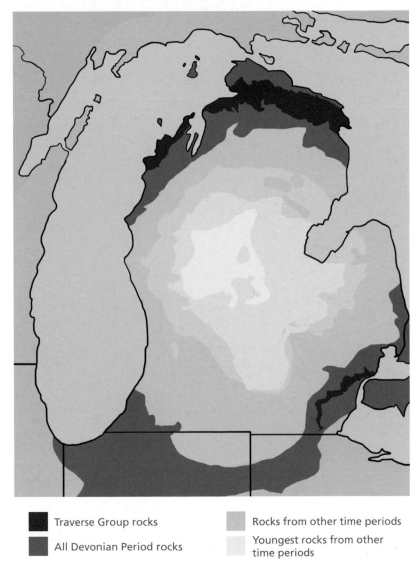

Traverse Group rocks

All Devonian Period rocks

Rocks from other time periods

Youngest rocks from other time periods

PETOSKEY STONE DISTRIBUTION

Much of the upper portion of southern Michigan's bedrock originates from the Devonian and is rich with fossil material. The Traverse Group rock layers, where Petoskey stones are found, are present primarily in the northernmost part of the Lower Peninsula in a band that stretches from Lake Michigan to Lake Huron. Though the entire group contains Petoskey stone and specimens can be found throughout the uppermost portion of the Lower Peninsula, the rocks are most exposed in the Little Traverse Bay area, particularly near Petoskey and Charlevoix. In those areas, Petoskey stones are more highly concentrated, making local beaches, dunes, and gravel-rich areas prime hunting locations.

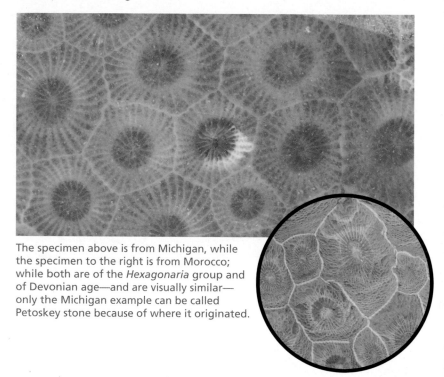

The specimen above is from Michigan, while the specimen to the right is from Morocco; while both are of the *Hexagonaria* group and of Devonian age—and are visually similar— only the Michigan example can be called Petoskey stone because of where it originated.

HOW DID PETOSKEY STONE GET ITS NAME?

Petoskey stone is named for the town of Petoskey, which is located at the mouth of Bear River on the shore of Little Traverse Bay, where the concentration of specimens has historically been at its highest. The town's unusual name is said to have come from the local legend of a French fur trader turned Odawa chief.

The legend, which has no doubt seen some embellishment over the past 140 years, states that a French fur trader by the name of Antoine Carre traveled to work in the burgeoning Great Lakes fur trade in the 1780s. Much trade was done with the indigenous peoples in the region, and Carre is said to have incorporated himself into their lifestyle and culture and married an Odawa princess. He is then said to have been accepted so fully into the local tribe that he was eventually made Chief. His tribal name became Neatooshing, and in the spring of 1787, his princess wife gave birth to a son. In the morning light, he declared his newborn son's name "Petosegay" (or some variant of the word), which translates roughly to "rays of sunlight." Throughout his life, Petosegay rose to prominence as an Odawa chief and successful fur trader, and not long before his death, a town was erected in his name in 1873. The spelling was anglicized to the modern-day "Petoskey."

While romantic, the veracity of the legend is up for debate. Official records claim that Reverend Andrew Porter, a Presbyterian missionary, founded the village of Bear River in 1852, which was later renamed Petoskey in 1873, in Petosegay's honor, after the Pennsylvania Railroad had reached the town and solidified its place as a center for trade. It is unclear what actual role Petosegay may have had in the founding of the town, but what we do know is that it was named for him and his tribe; the tribe's descendants still live in the Petoskey area today. A consistent feature in all tellings of the legend is the meaning of the word Petosegay, a word that certainly proved to be apt for Petoskey stone. Like the sun's rays, the septa seen in Petoskey stone's preserved coral are best described as ray-like, radiating outward from their central point.

VISIT PETOSKEY

Today, the town of Petoskey is a resort community that enjoys ample tourism, largely focused on the area's natural beauty, including the beaches where Petoskey stones can be found. Historic downtown Petoskey, called the Gaslight District, is a hub for many of the smaller communities in northwestern Michigan, so Petoskey is still an important center for regional trade.

Tourists and collectors snap up every available Petoskey stone find as wind and waves reveal them in and near Petoskey. However, that is not to say Petoskey stones are rare—on the contrary, Petoskey stone is still abundant throughout the Little Traverse Bay region. But savvy collectors know that they'll have better luck on the beaches less frequented, particularly those on the southern shore, between Charlevoix and Petoskey. If you don't have luck finding one, you can always visit one of the many shops in the area selling Petoskeys.

Specimen courtesy of Dean Montour

FINDING AND IDENTIFYING PETOSKEY STONES

After learning the history of Petoskey stones and how they came to be, you are probably wondering how and where you can find your own ancient piece of Michigan. Luckily for collectors, finding Petoskey stones is not difficult, as they are widely available to persistent visitors of the northwestern and northeastern shores of Michigan's Lower Peninsula. Identification can be trickier, however. With several very similar kinds of fossilized coral also present in the region, it can take a lot of research and perhaps some guesswork to tell them apart. In this section, we'll help you distinguish "true" Petoskey stones from some lookalikes and tell you where you can find them and other aquatic fossils.

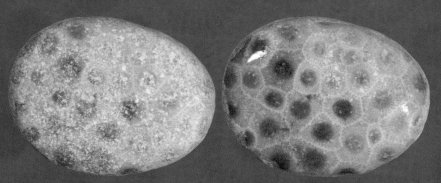

Wetting limestone will help bring out the Petoskey pattern, if present.

HOW DO I FIND PETOSKEY STONES?

Part of the reason Petoskey stone is so popular is because it's accessible for almost everyone to find. Finding Petoskey stone does not require any special skills or equipment, simply the desire to venture to natural spaces and turn over every pebble at your feet.

Because it's a variety of limestone, Petoskey stone can be tricky to spot among the multitude of other limestone and shale pebbles prevalent in the region, especially when the stones are dry. As weathering dulls the details of Petoskey stone, dry specimens are disguised as a chalky brown rock. For this reason, **if you suspect a stone is a Petoskey stone, get it wet**; water will help bring out the contrast in the fossil patterns, if present.

Petoskey stones are often found right on the beach, and nothing more than a sharp eye is needed to find them.

Little Traverse Bay

A small, nicely patterned Petoskey stone

51

COLLECTING RULES AND REGULATIONS

Before you start collecting Petoskey stones (or any other natural material), you need to be aware of the collecting rules and regulations where you are. First of all, collecting in national parks is always prohibited, and many state parks also forbid any collecting. One notable exception is Petoskey State Park, just outside Petoskey, where collecting is **allowed**. In other state parks, be sure to check the rules and regulations. Of course, you also cannot collect on private property if you don't have permission from the landowner.

And if you're collecting Petoskey stones on public land, the state of Michigan imposes strict limits on how much you can collect each year. An individual can only collect 25 pounds a year from public land. This means that you could collect many small specimens or just a few large ones, but once you've reached your limit, please respect these laws and leave your next finds behind.

If you happen to come across a large specimen that exceeds 25 pounds, you unfortunately need to leave it where you found it. In exceptional cases, when a very large piece in excess of 100 pounds or more is discovered, you're encouraged to report it to local officials. Those who have tried to take home such large discoveries are typically met with confiscation and fines.

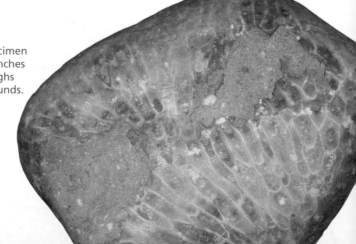

This large specimen measures 10 inches wide and weighs around 10 pounds.

A great specimen showing nearly perfect hexagonal structures as well as some attached reef material (dark brown)

Collecting Petoskey stones on Little Traverse Bay or any other area beach can be a wonderful excuse to go out and enjoy the scenery.

WHERE TO FIND PETOSKEY STONES

Thanks to the glaciers, specimens are not only widespread across the northern portions of Michigan's southern peninsula, but they are abundant as well. Many beaches, dunes, and rivers along Lake Michigan and Lake Huron, as well as some inland gravel exposures and dig sites, yield specimens of *H. percarinata* and other *Hexagonaria* fossils.

Areas that see lots of weathering and erosion, such as beaches and rivers, will produce the well-worn, rounded specimens that we see in abundance. More protected or more inland sites, such as gravel pits or other exposures of glacial gravel and sand, can yield more rough, ragged, sculptural specimens that better reflect the coral as it originally grew. The biggest difference between these two kinds of locales, however, is accessibility. Many of the best inland spots for hunting Petoskey stone are privately owned and cannot be visited without permission. In addition, the number of specimens they may produce can be sparing. Most of the best beaches, however, are public, and in many places, including Petoskey State Park, collecting is actually encouraged. And thanks to wave-action, specimens are plentiful and replenished often.

When hunting for Petoskey stones, little equipment or knowledge is necessary to make a discovery. If you're in the Little Traverse Bay area, all it takes is some sharp eyes and patience (although a bottle of water will help). The best way to begin your search is by checking virtually any pebbles or masses of limestone for fossil features. The distinctive honeycomb patterning is Petoskey stone's key defining trait, and wetting stones is a simple way of aiding identification. You may also find other fossils in the limestone.

Rougher, more finely detailed Petoskey stones like this are mostly, but not always, found in more inland areas.

Sandy gravel is a typical setting in which to find Petoskey stones.

The primary collecting areas in the northernmost portion of Southern Michigan

Key Petoskey stone collecting areas

Traverse Group rocks (fossil-bearing)

Other Devonian rocks

Other rocks

A PRACTICAL GUIDE TO PETOSKEY STONE (*HEXAGONARIA PERCARINATA*)

WHERE TO FIND IT (ENVIRONMENTS): Primarily on lakeshores, but also along riverbanks, road cuts, gravel pits, and any other gravel exposures in the northernmost portion of the Lower Peninsula. The entire Little Traverse Bay area and the vicinity of Alpena are best known for their fossil-rich beaches.

COLOR: Usually mottled in shades of gray to tan or brown, sometimes with a yellow tint. Rarely pink to red.

HARDNESS: Fairly soft; a pocket knife or piece of glass will easily scratch Petoskey stone and any other variety of limestone.

SPECIMEN SIZE: Specimens can vary greatly in size, but in general, beach finds are fist-sized or smaller.

DESCRIPTION: Generally light-colored limestone pebbles with conspicuous tessellated patterns of mostly hexagonal, or six-sided, shapes. Each hexagon is cell-like in structure, with a central point and radiating segments extending outward to the edges. Beach finds are usually rounded by water and may not show the honeycomb-like pattern until wetted. Finds from farther inland may be rougher, more ragged and irregularly shaped, sometimes with pronounced sculptural, cup-shaped hexagons and gauze-like texture. When worn or broken lengthwise, an elongated, parallel pattern, is instead visible.

IDENTIFYING DETAILS: *Hexagonaria percarinata*, the specific species of rugose coral found in "true" Petoskey stone, can be identified from other *Hexagonaria* species by close observation of its hexagonal pattern. Each hexagonal cell will average 10 mm across in size and contain 38 to 40 septa (the ray-like segments in each hexagon). Polishing a specimen will aid in the observation of these fine details.

Some typical beach-worn Petoskey stones as they were found

IDENTIFICATION CHECKLIST

1. Ensure your specimen is limestone. Petoskey stones are not found as any other type of rock.

2. Determine whether any fossil coral patterning is present. Wetting your specimen will help.

3. If a pattern is present, determine if the shapes are generally hexagonal.

4. If possible, measure size and count septa in each hexagon to determine if your specimen may be *H. percarinata* or another coral.

5. It is always a good idea to label each of your discoveries with the location where it was found.

Always be certain whether or not you are allowed to collect at your location.

While this seemingly dull piece of limestone may show slight indications that it contains coral fossils, it isn't very apparent until after it has been wetted.

PETOSKEY STONE COLORS

Virtually every specimen of Petoskey stone you'll find will be some shade of gray, brown, yellow, or tan, but some rarer hues do exist. Some exhibit reds or pinks, thought to be caused by iron impurities, and these are rare and highly sought-after by collectors. Some of these rarities contain "ribbons" of reddish color that follow the boundaries of the corallite walls, while others may have a pale pink coloration throughout the entire specimen. Particularly vivid specimens can be quite valuable among certain hobbyists.

Sometimes a Petoskey stone may appear to be green or greenish-yellow—this is almost invariably the result of recent algae or other plant growth. Since limestone is very porous, microorganisms can easily use its surface to anchor and grow. Most of the time, this coloration will fade after drying out, and it can usually be removed completely by washing.

A Petoskey stone with hints of pink-red coloration, shown wet

This rough specimen shows an orange-pink coloration that is slightly more common than the deep pinks or reds.

Specimen courtesy of Bob Wright

A rare Petoskey stone with red "ribbons" that follow the corallite walls, shown dry (above) and wet (below)

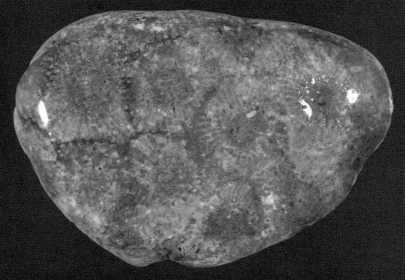

EXPECT A LITTLE VARIATION

While the corals in Petoskey stone are six-sided by definition, not all corallites exhibit a perfect shape. Many are skewed or lopsided, and some that are very "crowded" in the pattern may only have five, four, or three prominently visible sides. This is typical and indicative of the tightly compact nature of the corallites. Even so, the majority of corallites in a specimen will exhibit six sides. If you have found a piece of coral-bearing limestone that shows no six-sided corallites at all, then you likely have found fossilized coral of a different genus entirely, and it's therefore not a true Petoskey stone.

The specimen to the left has essentially perfect hexagonal patterning, but in the specimen above, note how many of the corallites appear to have as few as five sides and as many as eight. This is due to the crowded nature of corallite growth.

OTHER VIEWS

While the tightly packed hexagonal pattern is the most popular and well-known appearance for Petoskey stone, it is important to remember that Petoskey corals were three-dimensional structures. *Hexagonaria percarinata* developed as dome-like clusters of corallites; the top side of the dome was covered in individual polyps. As it grew upward and outward, increasing the dome size, the individual parallel corallite structures also grew upward and became longer as the polyps added to their calcite skeleton. Therefore, when viewed from the side, the long, parallel corallite structures are seen instead of the hexagonal shapes. This is an interesting way to view and study the corals in Petoskey stone, as it offers a view inside the corallites.

These side views of corallites show their elongated internal growth structure.

FROM THE UNDERSIDE

Finds showing the top and sides of *H. percarinata* are very common, but the underside is another matter. Since domes, often called heads, of *H. percarinata* could grow to enormous sizes, many were not preserved entirely—especially after the glaciers plowed through the region. Most broke up and became the smaller pieces we find on the beaches today. Yet some whole dome structures did survive, and some are surprisingly small. While tricky to identify, a feature to look for is something that looks like a "stem." You can also look for an obvious dome-like structure, but many water-rounded specimens have a similar look.

Many *H. percarinata* domes were somewhat mushroom-like in shape, with a short base from which the coral colony anchored itself and began growing. This unique structure can be conspicuous with ring-like layers or bands around it. In most water-worn specimens, this structure's presence is nearly the only way you can tell if you've found a whole dome or not.

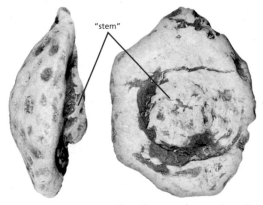

"stem"

The top side view of this specimen shows a typical Petoskey stone pattern.

But views underneath and of the side reveal the "stem" at its base and the circular structures around it.

These three specimens are all photographed from the bottom, or underside, of coral domes, showing the often unusual ring-like patterns.

This specimen was polished along the top of the corallites and down along the side, illustrating the corallites' elongated structure and how the hexagonal shapes extend deeper into the stone.

ATTEMPTING TO IDENTIFY A FIND CONCLUSIVELY

Once you've determined that you've discovered a *Hexagonaria* fossil, you'll need to distinguish it from other similar fossils. Other corals, both from the *Hexagonaria* genus and from other coral groups, are present in the Traverse Group rocks, and many of them look very similar to *H. percarinata*. Be warned: Differentiating them can be difficult, to say the least.

To start out identifying your find, you need to know a little about how scientists classify animals and plants. When scientists describe an organism, they refer to a genus name and a species name. A genus name is similar to a surname, as it indicates that all members of that genus are related. Along those same lines, a species name is similar to an individual name.

The coral that produced the Petoskey stone was *Hexagonaria percarinata*; its species name is *percarinata* and its genus, or most closely related group, is *Hexagonaria*. The other six-sided corals in the *Hexagonaria* genus are very closely related to *percarinata* and share a very similar appearance. And, fortunately for fossil enthusiasts, they are all present in the Traverse Group rocks.

Hexagonaria percarinata was designated as the "true" Petoskey stone because it is by far the most abundant and common member of *Hexagonaria* in the area's rocks. But nine other *Hexagonaria* species have been identified in the Traverse Group's limestone. For collectors, this means that any given specimen of Petoskey stone probably contains *H. percarinata*, but not necessarily. All specimens are informally considered Petoskey stone, but those with the knowledge and patience may be able to find scarcer species in the rocks.

THE TECHNICAL DETAILS

Die-hard Petoskey collectors, however, insist that "true" Petoskey stones contain only *percarinata*. If you're deadset on conclusively identifying your find as *H. percarinata*, it's a tricky task. To identify an individual species, collectors need to measure the average size of the corallites (in millimeters) and count the average number of septa in each individual corallite.

The list below details the 10 species of *Hexagonaria* present in Michigan's Traverse Group of Devonian rocks and their typical corallite sizes and number of septa. **Note:** There are overlaps between multiple species; for this reason, *Hexagonaria* species are difficult to identify definitively without in-depth research. In addition, weathered or water-rounded specimens can be difficult to examine for detail. Other distinguishing characteristics (e.g. septa wall thickness) are even more difficult for novices to observe and research. With that said, the information below will at least give you a fighting chance to identify your find by species.

Hexagonaria (abbreviated as *H.*) species in Michigan's Devonian rocks, organized by average corallite size (in millimeters)

H. alpenensis	2–6 mm	26–28 septa
H. potterensis	4–15 mm	36–38 septa
H. fusiformis	6–10 mm	34–38 septa
H. subcarinata	8–12 mm	30–34 septa
H. attenuate	8–14 mm	30–38 septa
H. percarinata	10 mm	38–40 septa
H. anna	12–16 mm	32–44 septa
H. profunda	13–15 mm	38–42 septa
H. cristata	16–20 mm	36–44 septa
H. mirabilis	NA	NA

None of the various *Hexagonaria* species in Michigan are more common than *Hexagonaria percarinata*, the "true" Petoskey stone coral. Therefore, when finding a specimen, it will be safe to assume that it probably contains the species determined to be the one in the state stone. But if you want to confirm this, you can measure the corallites and count their septa, as discussed on the previous page.

The corallite size of *Hexagonaria* species has been documented in detail and is key to distinguishing them. When looking at the hexagons in a Petoskey stone, you'll notice that most are close to the same size. By carefully measuring several corallites at their widest point in millimeters and then averaging them, you can begin to narrow down which species you've found. Then, with the help of good magnification, you can count the septa—the ray-like segments—in several corallites and find the average of that number too. (This is most easily done in polished specimens with very clear patterns.) With these two pieces of information, consult the chart on the previous page to see which species you may have.

H. percarinata has 10 mm corallites with 38 to 40 septa, which may seem very exact and distinctive, but because of the similarities and overlap in these traits with other species' ranges, *H. percarinata* actually shares its measurements with at least two other species. But aside from the abundance of *H. percarinata*, it also has a very regular, even corallite size with little variation, both of which are traits that can further help identify it.

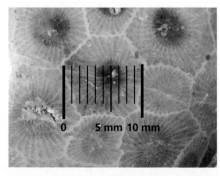

Specimen courtesy of Dean Montour

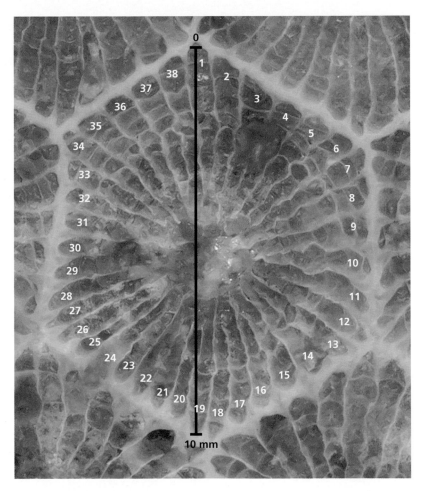

This close-up shows an idealized *H. percarinata* corallite. It is almost exactly 10 mm at its widest point and contains exactly 38 septa, numbered above. This specimen can be safely assumed to be a "true" Petoskey stone.

PETOSKEY LOOKALIKES; OTHER *HEXAGONARIA* SPECIES

With no fewer than 10 *Hexagonaria* species present in the Traverse Group limestones, all with very similar features, it can be daunting for novices trying to find *H. percarinata*. The differences between *Hexagonaria* species can be extremely subtle—so subtle that the majority of collectors won't notice them—but with some time and practice, you can get better at spotting *H. percarinata* as well as several lookalike species.

Realistically, it won't be possible for anyone without significant knowledge of coral biology or outside of a lab setting to identify all 10 present species. But that shouldn't stop you from trying. Being able to apply your knowledge to assign even tentative species names to your specimens makes for a much more interesting and fulfilling collection. Thankfully, narrowing a specimen down to two or three possible species is relatively easy, once you know to measure average corallite size and count their septa. From there, though, only additional research will help pinpoint a specific coral.

This specimen has corallites that average 7–8 mm with 34 septa and is likely *H. subcarinata* or *H. attenuate*.

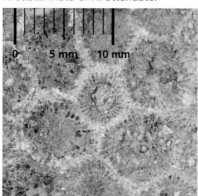

This Charlevoix stone contains very small corallites of less than 5 mm and is likely *H. alpenensis*.

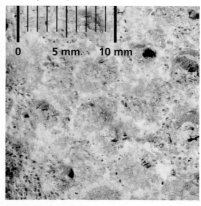

Farther east, at less frequented exposures of Devonian rocks, there is another species, *Hexagonaria alpenensis*, that is less commonly seen but can be easy to identify. Named for the Alpena Limestone, which is another rock layer in the Traverse Group, *H. alpenensis* is the smallest species of *Hexagonaria*, with corallites measuring less than 6 mm, making it very easy to visually tell them apart from *percarinata*'s much larger corallites. Similarly, another variety of Petoskey stone known locally as Charlevoix stone also contains a different member of *Hexagonaria* with smaller corallites, but it is less distinctive than *H. alpenensis*. However, many misidentified or misunderstood "Charlevoix Stones" actually contain coral of another genus entirely, known as *Favosites* (page 74).

The specimen below has a mixture of different corallite sizes, ranging from about 6 mm to around 10 mm. It may be *H. fusiformis*.

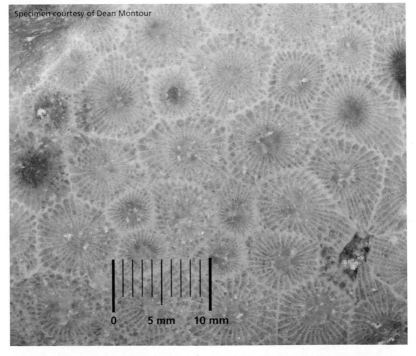

Specimen courtesy of Dean Montour

0   5 mm   10 mm

OTHER THINGS YOU MIGHT FIND: HORN CORAL

The Devonian seas saw a dramatic increase in marine life, and reefs from the period were home to many species, dozens of which can be found in the Traverse Group's rocks. It may require some in-depth research to definitively identify any of them, but arriving at a ballpark identification isn't generally very difficult.

Solitary rugose corals, particularly those we call horn coral, were related to *Hexagonaria* corals but did not live in colonies, instead forming solitary tube-like structures. These "horns" are generally simple to identify, as their elongated, often curved shape is distinctive. In addition, their corallites are usually wider than those found in *Hexagonaria* fossils. Horn corals also sometimes boast gauze- or fabric-like textures or ring-like bands of color.

Horn corals are typically found loose but may be embedded in limestone as well; even within rock, their size and shape make them easy to identify.

Specimen courtesy of Dean Montour

These are all examples of horn coral from the Petoskey region.

A large horn coral still embedded in limestone

OTHER THINGS YOU MIGHT FIND: FAVOSITES

Hexagonaria is not the only genus of colonial coral present in the Traverse Group rocks. *Favosites* is another long-extinct genus of coral that grew in tightly packed colonies of six-sided corallites with tessellated honeycomb-like patterns. They lived in Earth's seas much longer than any species of *Hexagonaria,* and several species are present in Michigan's Devonian rocks. They are generally under-studied, and telling different *Favosites* species apart from each other can be exceedingly difficult. But, once you know what to look for, distinguishing them from *Hexagonaria* corals and Petoskey stone is easier, despite their visual similarities.

The first thing you'll notice about *Favosites* is that their corallites tend to be much smaller and more delicate in appearance. With thin corallite walls, minute septa, and virtually absent columellae, *Favosites* are simpler-looking than *Hexagonaria* fossils. In addition, the smaller corallites are more parallel, longer, and more plentiful, and also more uniform in size. In many *Hexagonaria* specimens, some smaller corallites may be seen packed between larger ones, but *Favosites* tend to have a more evenly spaced structure. Careful examination of individual corallites will help you tell the specimen apart from those of Petoskey stone.

Favosites tend to have much smaller corallites and more of them, tightly packed in dense groups.

Biologically, *Favosites* differed from *Hexagonaria* in that their corallite walls had small perforations that enabled them to transfer nutrients between polyps without the need for connective coenosarc. The average collector will not be able to observe such a tiny detail, but particularly well-preserved specimens may exhibit these tiny structures under strong magnification.

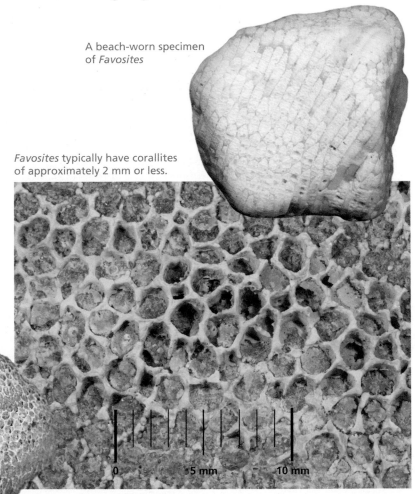

A beach-worn specimen of *Favosites*

Favosites typically have corallites of approximately 2 mm or less.

0   5 mm   10 mm

OTHER THINGS YOU MIGHT FIND: HALYSITES

Halysites are another coral you may find in the region. Also known as chain corals, these rare corals form in chain-like growths, but they are not found in Devonian-age rocks. Instead, *Halysites* largely went extinct in the Silurian Period, just before the Devonian, and they are native to the older rocks of the southern portion of Michigan's Upper Peninsula. They are occasionally found in the Little Traverse Bay area and other Petoskey stone-rich localities thanks to the incredible power of the glaciers of the most recent ice age.

The older *Halysites* chain corals tend to be more fragile than Petoskey stones, with more-delicate features that are easily weathered away. For this reason, they are scarce and highly prized by local beachgoers. Aside from being corals, they share little in common with *Hexagonaria* species. You may sometimes see chain corals referred to as "parasitic," but this is a misnomer derived from their occasional presence with or upon other ancient corals.

A water-worn stone containing clear *Halysites* chains

Specimen courtesy of Dean Montour

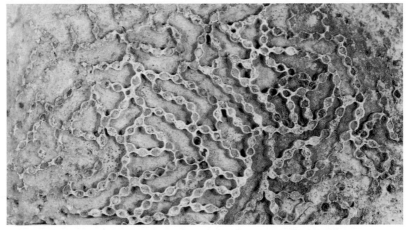

This unusually fine specimen illustrates the delicate chain-like structure of *Halysites*. Though it was found near Petoskey, it originated much farther north.

UNCERTAIN DISCOVERIES

Other corals found in southern Michigan may be less distinctive and harder to identify. Coral species, whether solitary or colonial, often have vague branching shapes, gauze-like textures and patterns, and discs with radiating coloration. If you want to identify your other finds, see page 95 for a list of additional resources. You will likely pick up some strange fossils that you won't be able to identify with books alone. For these, contacting a local club or museum to talk with a professional may be your best chance to figure out just what you've found.

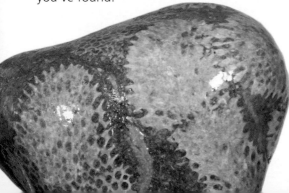

While it may resemble coral, less common fossils like this bryozoan can stump novices.

OTHER COMMON FOSSIL FINDS

Since the Traverse Group rocks were once part of a vibrant ocean ecosystem, it should come as no surprise that other fossilized reef-dwelling species can be found on the same beaches and gravel pits as Petoskey stone. Snails, bryozoans, shellfish, and even petrified wood are all possible finds. These fossils will likely appear in your hunts for Petoskey stones.

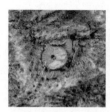  **Crinoids**—Nicknamed "sea lilies," these were plant-like animals that anchored themselves on the seafloor with a long, tubular stem. When they died, the disc-shaped segments of the stem frequently separated and scattered into the reef. These appear as small circles in limestone, often with a hollow or a star-shaped center.

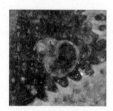 **Gastropods**—Typically snails or snail-like shells. Remains of these hard-shelled critters will appear as small coils or something like a cluster or chain of circular shapes. When seen in cross section, as is typical in broken reef limestone, they may appear as small outlined circular or "scoop"-shaped structures.

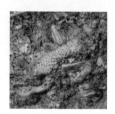 **Bryozoans**—Another group of colonial animals that share a skeleton, bryozoans are nicknamed "moss animals" due to the particularly plant-like appearance of their colonies. Still in today's oceans, the individuals in a bryozoan colony are all clones that feed by filtering water through their bodies. Fossils appear very much like corals but have a finer appearance with countless tiny pores in their surfaces, and small fragments can almost appear sponge-like.

Shell Fragments—In many chunks of reef limestone, you'll see find lots of small, curving or fan-like shapes—these are typically fragments of shells from gastropods (such as snails), bivalves (clam-like animals), or brachiopods. While whole specimens do exist in the area rocks, most will show up as little pieces, sometimes attached to Petoskey stones.

Petrified Wood—While certainly not a reef fossil, it is possible to find small pieces of fossilized wood, often called petrified wood, on Michigan's beaches. This material, which is younger than the Devonian rocks, was deposited by the glaciers and mostly originated farther north. Identification can be tricky, since other rock fragments can look wood-like, but checking for wood grain and bark-like features will help.

Trilobites—These unusual looking animals lived on seafloors for nearly 300 million years but are all extinct today. Their hard armored bodies had a bulbous head and plate-like tail connected by rib-like segments between. These are fairly rare in Michigan's Devonian rocks, especially whole, but can be found in the area. These are often the "grand prize" for Michigan fossil collectors; these are found in shale, particularly around Alpena.

Trilobites from Michigan are rare and highly sought after. This specimen is actually from Ohio but is of Devonian age and belongs to a particular family of trilobite species that are found in the Traverse Group rocks. Michigan trilobite specimens will look very similar to this.

WHAT TO DO WITH YOUR FINDS

Once you've begun collecting Petoskey stones, there comes the question of what to do with them next. Labeling them with the location where they were collected and displaying them is the classic option, but what if you want to polish your specimen or turn it into jewelry? In this section, we'll look at a few of the things you can do with your discoveries to make them more presentable and discuss what they may be worth.

This group of rough, inexpensive stones doesn't show much patterning in their rough state, and many would benefit from polishing.

Specimen courtesy of Dean Montour

Polishing helped reveal this otherwise pale patterning. Polishing also made it easy to see that the corallite sizes differ greatly throughout the specimen, which indicates that it may be a different species and not *H. percarinata*; this detail probably wouldn't have been observable in a rough specimen.

WHAT CAN I DO WITH IT?

Once you find a Petoskey stone, one of the most popular ways to enjoy it is by polishing it. Polishing makes Petoskey stone's distinctive patterning much more apparent, bringing out contrast and definition within the shapes without the need to wet your specimens.

Most rock and mineral polishing is done with lapidary equipment, such as diamond-bearing grinders and tumblers, but because it's made of limestone, Petoskey stone doesn't require expensive equipment to shape and polish it. And unlike other relatively soft materials, which often don't take a good polish and look somewhat dull, limestone's calcite-dominant composition and relatively densely packed grains enable it to take a fine polish. This makes polished Petoskey stone specimens perennially popular.

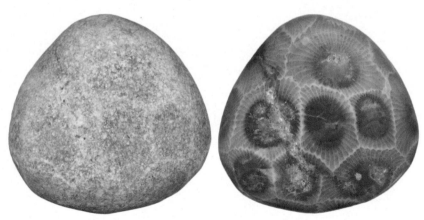

A common pebble before polishing (left) and after (right). Note the crack-like structure that was made visible by polishing. This was where the stone was once broken long ago but has since been filled by quartz that crystallized, thereby "healing" the stone back together.

This specimen, from a rock dealer's "junk bin," actually had a lot of detail to show off once it was polished. (Rough left, polished right)

HOW TO POLISH PETOSKEY STONE

Getting nicely polished Petoskey stones doesn't require expensive equipment, and great results can be achieved by hand with nothing more than sandpaper. Hand-sanding Petoskey stones works best on specimens that have already been shaped and rounded by wind and waves, as their smooth surfaces will be easier to work with. Dry or wet sandpaper may be used, but whatever your choice, the stone itself should be wetted before sanding. Begin by starting with fairly coarse sandpaper—from 100 to 220 grit—and grind down the surface of a specimen until it is uniform, being careful to go over any deep scratches or pits extensively. This first step is the most important, as it will determine how well-shaped and smoothed the stone will be when you're finished, so take your time—it will be a slow process, taking anywhere from 30 minutes to a couple of hours, depending on the size of your stone and how rough it was to begin with. Rinse the stone often to eliminate coarse grains that may continually scratch the stone.

Once you feel that the stone is sufficiently sanded, proceed to use finer and finer sandpaper—400 and then 600 grit—to repeat the process several times. As you progress to finer sandpaper, the overall appearance of the sanded area should become smoother and details within the stone should become clearer. You can perform as many intermediate steps as you wish, but in general, the higher the final grit number you use, the better the final polish. To achieve the finest possible polish, you can even use a polishing compound, which contains an ultra-fine grit. These powders or pastes are usually applied to a piece of heavy fabric, like velvet.

Horn corals polish up just as well as Petoskey stones.

Specimen courtesy of
Dean Montour

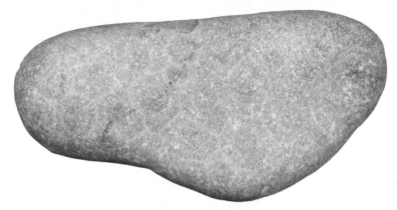

Polishing can reveal interesting details that would remain hidden otherwise. This specimen shows that it was once broken when still within the Earth, and the pressure of burial moved the two broken pieces slightly. Later, when a calcite solution filled the crack and crystallized, "healing" the stone, this slight jog in the pattern was preserved.

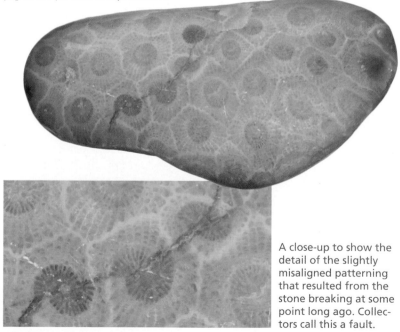

A close-up to show the detail of the slightly misaligned patterning that resulted from the stone breaking at some point long ago. Collectors call this a fault.

OTHER POLISHING OPTIONS

The entire hand-polishing process can take hours, depending on how fine a result you want. But there are other ways to polish Petoskey stones. Rock tumblers, which have a rotating drum into which you place stones, water, and polishing grit, can yield specimens that are nicely polished on all sides, but they can take around a week and using them can be messy. In addition, tumblers also remove a lot of material from each specimen, resulting in smaller stones. But they are affordable and automated, requiring only periodic maintenance throughout the process. Other motorized equipment, such as wheel-based diamond gemstone polishers, can make the process much faster but require considerable investment and expertise. If polishing your finds yourself isn't an option, local lapidary professionals or hobbyists could do the work for you. Local rock clubs or online groups can put you in touch with someone willing to work with you. Whatever method you use to polish Petoskey stones, patience is key; even though the stone is soft and easy to work with, it's easy to scratch, break, or grind away completely.

Some collectors avoid polishing altogether but still achieve that "wet" look by coating specimens in a clear coating of some sort, typically an epoxy or resin. These thick, clear fluids can be used to coat a stone and, once dried, form a permanent coating that can give specimens a richer, more contrasting, "always wet" appearance. However, as limestone is porous and soft, results can be mixed, and this method of treatment should only be used by experienced hobbyists.

This rough piece of Petoskey stone has been coated in a light silicone spray, enhancing its contrast slightly without excess mess.

Jewelry

Petoskey stone is frequently used in jewelry and stone carvings, thanks to its interesting patterns and how easy it is to work with. Knowledgeable lapidaries can cut Petoskey stone at specific angles to achieve the pattern of their choosing or to accentuate features of a specimen. All manner of Petoskey stone jewelry and other wearables and trinkets are abundant and widely available, especially in the Petoskey area.

Petoskey in Aquariums

Some aquarium hobbyists may be tempted to put Petoskey stones or other fossil corals in their fish tanks. This may be fine in the short term, but it could be problematic in the long term. Because limestone consists of calcite, it dissolves easily. As a fish tank's water becomes less fresh, accumulating waste and other organic by-products, the water can become more acidic, thereby dissolving the limestone. The process may be slow, especially if the fish tank is well-kept, but eventual erosion of the specimen will likely occur. More importantly, this can change the pH balance of the water, which can be harmful to fish.

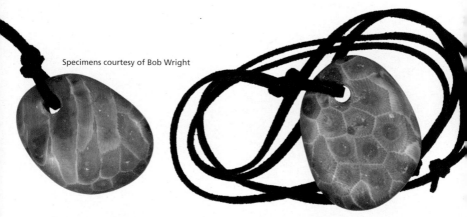

Specimens courtesy of Bob Wright

These lightly polished Petoskey stone beads are some popular ways to wear the stone while retaining its more natural appearance.

WHAT IS MY PETOSKEY STONE WORTH?

While collecting Petoskey stones is an activity best appreciated for promoting an understanding of natural history, for many collectors it is undeniable that some specimens can be valuable. Once a specimen is collected, many hobbyists' next question is, "what is this worth?" The answer is, of course, that your specimen is worth whatever someone is willing to pay. This is because the valuation of Petoskey stones is very subjective. Size, clarity of pattern, color, and overall shape and condition of the stone are all factors that come into play. Stones that show interesting features, such as the underside of a coral dome or attached fossils of other animals, will fetch higher prices than more mundane samples. And popularity and location are something to keep in mind too. While specimens are common in the Little Traverse Bay region, they are also more popular there than anywhere else. Outside of Michigan, however, the value of small, average specimens plummets, while exceptional specimens retain their value.

In general, the most valuable specimens will be those that are fist-sized or larger with particularly bold, clear patterns and coral structure. Red or pink colorations are rare and will help increase a specimen's value dramatically. And polished specimens frequently fetch higher prices, but not always; sometimes a rough sample will have enough character and quality to stand on its own merit. In all cases, remember that the price you can expect to receive from a shop owner or professional collector will be less than the retail value of a stone. If your specimen is valued at $20 retail price, you can expect to receive half (or less) of that amount.

Most Petoskey stones are inexpensive, with small, average quality, unpolished specimens frequently fetching $1 or less. Finer, larger specimens often land in the range of $10 to $50, but only the finest, most interesting, and best-patterned pieces will reach the $100 price point and up. Don't be disheartened if your specimens aren't worth much; there are always more and better specimens out there to be found.

Average beach-worn specimens like these, which measure about 2–3 inches across, are frequently under $1 each in value and are often bought and sold in bulk quantities.

This larger, rough specimen, measuring about 8 inches across, may show more detail than a water-worn sample, but its ragged appearance keeps it from being very valuable, and it is worth under $20.

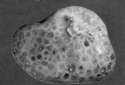

This polished specimen, measuring about 7 inches across, shows clear, bold patterning over most of the surface of the stone. While not a museum-grade specimen, it is still a desirable example that any collector would be proud to find; it is worth $25–$50.

A FAVORITE FIND

Petoskey stone is a perennial favorite in Michigan, and it is seen in collections and museums all over the world. Undoubtedly Michigan's most famous fossil, and certainly one of the United States' as well, the popularity of Petoskey stone is a testament not only to the charm of its enigmatic patterns, but also to how easy it is to find, collect, and polish. But now when you find your own specimen, you can also appreciate something else: the 400 million years that have passed since its tiny little polyps dutifully began building the coral reef they called home.

Petoskey stone is a relic from a critical period for life on Earth and helped form the backbone of the ancient seas: the reefs that harbored life and protected our planet's early animals. While *percarinata* and the other *Hexagonaria* species were ultimately unable to withstand the trials of time, they did thrive for approximately 57 million years and were a successful group of species by any definition. But within their hexagonal patterns and layered structures is also an opportunity to better understand the challenges our oceans have faced and continue to struggle with today. Corals, then and now, are an indicator of the health of our waters, and as coral bleaching spreads through our modern reefs we must be wary; the events of the Late Devonian extinction may be in the distant past, but nothing precludes them from occurring again.

Discovering any fossil is an exciting experience, but if you know the history behind it, collecting can be even more enjoyable. This is something that many Michiganders and visitors to the Petoskey area have long understood, and it's something that will likely persist in the future as well.

This Petoskey stone shows large overgrowths of coarsely crystallized calcite that mimic the hexagonal structure.

Specimen courtesy of Alex Fagotti

GLOSSARY

Algae—A general name for nonflowering plants, which can include single-celled organisms as well as seaweeds. They produce energy via photosynthesis but lack stems, roots, and leaves and are typically aquatic.

Aragonite—A mineral form of calcium carbonate, often forming elongated, needle-like crystals. It is one of the primary forms of calcium carbonate produced by mollusks for their shells.

Calcite—The most common mineral form of calcium carbonate, often forming elongated six-sided pointed crystals. It is one of the primary forms of calcium carbonate produced by corals, foraminifera, and other sea life.

Calcium Carbonate—A chemical compound consisting of calcium, carbon, and oxygen with the chemical symbol $CaCO_3$. Depending on the conditions present, it can take the form of aragonite or calcite, which differ in crystal structure. Corals, mollusks, and other sea life produce it.

Class—A larger biological subdivision of animals, such as insects or mammals. Corals belong to the class Anthozoa, which also includes sea cucumbers and other marine organisms.

Coenosarc—The living tissue that coats a coral skeleton and connects polyps.

Columella—The central rod-like structure within a corallite that supports the septa and enables the polyp to grow upward.

Colonial coral—A coral species that develops as a tightly grouped mass of individual polyps that all contribute to a singular coral skeleton.

Coral—General name for the marine stony external skeletons produced by coral polyps.

Corallite—The cuplike calcium carbonate skeleton of a single coral polyp; many corallites make up a coral colony.

Crystal—A solid body with a repeating atomic structure formed when an element or chemical compound solidifies.

Devonian Period—A period of geologic time occurring between approximately 419 and 359 million years ago, characterized by diverse sea life and the rapid development of life on land.

Eutrophication—Excessive richness of nutrients in a body of water, usually caused by runoff from the land. It typically creates dense plant growth that results in a low-oxygen aquatic environment that suffocates animals.

Extinction event—An event or series of events causing catastrophic die-offs of plants and animals, resulting in numerous species going extinct.

Fossil—Remains of plants or animals that became mineralized and were preserved within rock.

Fossil record—The record of the history of life as preserved within rock layers; the totality of all fossils, which shows us when species lived and when they went extinct.

Fossilization—The process by which remains of plants and animals are turned to minerals within rocks, usually through chemical reactions or slow replacement by minerals.

Genus—A smaller biological subdivision that classifies more-closely related groups of animals, such as specific groups of related corals. Genera are labeled with a capitalized Latin name.

Glacial period—A period of time during an ice age that is characterized by significant southward glacier movement.

Glacier—A slow-moving mass of ice formed by the compaction of snow.

Hexagon—A shape with six sides; something with this shape is referred to as hexagonal.

Hexagonaria percarinata—The specific species of coral preserved in Petoskey stone.

Horn coral—see "Solitary Coral."

Ice age—A period of time defined by lower-than-average global temperatures and glacier formation and glacial activity.

Invertebrate—Animal without a spine.

Limestone—A common sedimentary rock formed primarily of calcite derived from ancient sea floor sediments, including shells and coral reef material.

Marine—Of or relating to oceans or seas.

Mineral—A naturally occurring chemical compound or native element that solidifies with a definite internal crystal structure.

Photosynthesis—The process by which plants and some other organisms use sunlight to produce food from carbon dioxide and water, generally producing oxygen as waste.

Polyp—A small tube-shaped marine organism, usually with a tentacle-ringed mouth, which may be solitary or live in a colony, and often produces a hard calcium carbonate external skeleton.

Quartz—The most common mineral, quartz consists of silica (silicon dioxide) and forms pointed six-sided crystals that are colorless to white. It is also a constituent of many rocks.

Reef—A stony ridge in a shallow marine environment that is formed by large, inter-grown formations of corals and other marine life. Reefs can be large and complex and are vital to the survival of most marine life.

Rock—A massive aggregate of mineral grains.

Rugose coral—An extinct group of both solitary and colonial corals characterized by ridged or wavy external surfaces, resembling wrinkles.

Sedimentary rock—Rock derived from sediments of minerals, animals, or other rocks that have since been cemented and solidified together.

Septa—The individual chambers of a corallite skeleton, typically appearing as ray-like formations stemming from the central columella. Singularly called a septum.

Solitary coral—Coral species that do not live in tightly packed colonies and instead develop as singular tube-like coral structures with a single polyp. Many are known as "horn corals" due to their typical long, tapered, curving, horn-like shape.

Species—A biological subdivision consisting of similar individuals capable of inter-breeding or exchanging genes. Every species belongs to a genus, which in turn belongs to a class.

Stony coral—Any species of coral that produces a hard, mineralized skeleton composed of calcium carbonate. Stony corals usually contribute to reefs.

Tessellation—A group of shapes that fit closely together, usually repeating and without gaps or overlaps.

Zooxanthellae—A single-celled organism that lives within coral polyps and produces energy from the polyps' waste, in turn producing oxygen for the polyp.

BIBLIOGRAPHY AND RECOMMENDED READING

Bates, Robert L., editor. *Dictionary of Geological Terms, 3rd Edition*. New York: Anchor Books, 1984.

Coenraads, Robert R. *Rocks & Fossils: A Visual Guide*. Buffalo: Firefly Books, 2005.

Heinrich, E. W., and Robinson, George. *Mineralogy of Michigan*. Houghton: Michigan Technical University, 2004.

Milstein, Randall L. "Middle Devonian Traverse Group in Charlevoix and Emmet counties, Michigan." Geological Society of America Centennial Field Guide, North-Central Section, 1987. www.michigan.gov/documents/deq/GIMDL-GSA87J_302415_7.pdf

Mueller, Bruce, and Wilde, William H. *Complete Guide to Petoskey Stones, The*. University of Michigan Press, 2004.

Stumm, Erwin C., and Tyler, John H. "Corals of the Traverse Group of Michigan Part XII, The Small-Celled Species of *Favosites* and *Emmonsia*." University of Michigan, 1964. https://deepblue.lib.umich.edu/bitstream/handle/2027.42/48382/ID227.pdf

Stumm, Erwin C. "Corals of the Traverse Group of Michigan, Part 13: *Hexagonaria*." University of Michigan, 1970. https://deepblue.lib.umich.edu/bitstream/handle/2027.42/48447/ID296.pdf

Vicary, Joel. *Over 100 Collecting Locations in Lower Michigan*. Self-published, 2007.

Walker, Cyril, and Ward, David. *Fossils*. New York: DK Smithsonian Handbooks, 2002.

Wilson, Steve E. "The Petoskey Stone." Michigan Department of Environmental Quality, Geological Survey Division. www.michigan.gov/documents/deq/ogs-gimdl-GGPS_263213_7.pdf

Zeitner, June Culp. *Midwest Gem, Fossil and Mineral Trails of the Great Lakes States*. Baldwin Park: Gem Guides Book Company, 1999.

ABOUT THE AUTHOR

Dan R. Lynch grew up in his parents' rock shop—Agate City, in Two Harbors, Minnesota—on the shores of Lake Superior, where he learned the nuances of rock, mineral, and fossil identification firsthand. He has always enjoyed sharing his knowledge, and after earning his degree in graphic design with emphasis on photography from the University of Minnesota Duluth, it seemed a natural choice to combine all of his interests by writing rock and mineral field guides. His father, Bob Lynch, a respected veteran of the Lake Superior agate-collecting community, helped him get his start in this endeavor, and now Dan enjoys helping amateurs "decode" the complexities of geology and mineralogy. Many of his books focus on specific favorites, such as Lake Superior agates or Petoskey stone, and he takes pride that they all feature his true-to-life photographs and easy-to-read text. He currently lives in Madison, Wisconsin, with his wife, Julie, where he works as an author, artist, and classical numismatist.